SCIENCE QUESTIONS

WHY DOES A BOAT FLOAT?

by Elizabeth Andrews

Cody Koala

An Imprint of Pop!
popbooksonline.com

abdobooks.com
Published by Pop!, a division of ABDO, PO Box 398166, Minneapolis, Minnesota 55439.

Printed in the United States of America, North Mankato, Minnesota

102021
012022

THIS BOOK CONTAINS RECYCLED MATERIALS

Cover Photo: Shutterstock Images
Interior Photos: Shutterstock Images, 1, 5 (top, bottom left, bottom center), 7, 8–9, 11, 12 (top, bottom), 15, 17 (top, bottom left, bottom center), 18–19, 21

Editor: Grace Hansen
Series Designer: Laura Graphenteen

Library of Congress Control Number: 2021942249

Publisher's Cataloging-in-Publication Data
Names: Andrews, Elizabeth, author.
Title: Why does a boat float? / by Elizabeth Andrews
Description: Minneapolis, Minnesota : Pop!, 2022 | Series: Science questions | Includes online resources and index.
Identifiers: ISBN 9781098241124 (lib. bdg.) | ISBN 9781098241827 (ebook)
Subjects: LCSH: Boats--Juvenile literature. | Buoyant ascent (Hydrodynamics)--Juvenile literature. | Density--Juvenile literature. | Children's questions and answers--Juvenile literature.
Classification: DDC 500--dc23

Hello! My name is

Cody Koala

Pop open this book and you'll find QR codes like this one, loaded with information, so you can learn even more!

Scan this code* and others like it while you read, or visit the website below to make this book pop.

popbooksonline.com/boat-float

*Scanning QR codes requires a web-enabled smart device with a QR code reader app and a camera.

Table of Contents

Chapter 1

What Floats That Boat?

Have you ever wondered why something as big as a cruise ship is able to stay on top of the water? There are even boats big enough for airplanes to take off from and land on. How do these boats float?

cruise

The oldest boat ever discovered is nearly 10,000 years old.

shipping

military

Watch a video here!

Chapter 2

Buoyancy

Boats can float because they are **buoyant**. Things are buoyant because different forces act on each other. The upward force of the water pushes against the downward force of the boat.

boat force pushing down
buoyant force pushing up
Learn more here!

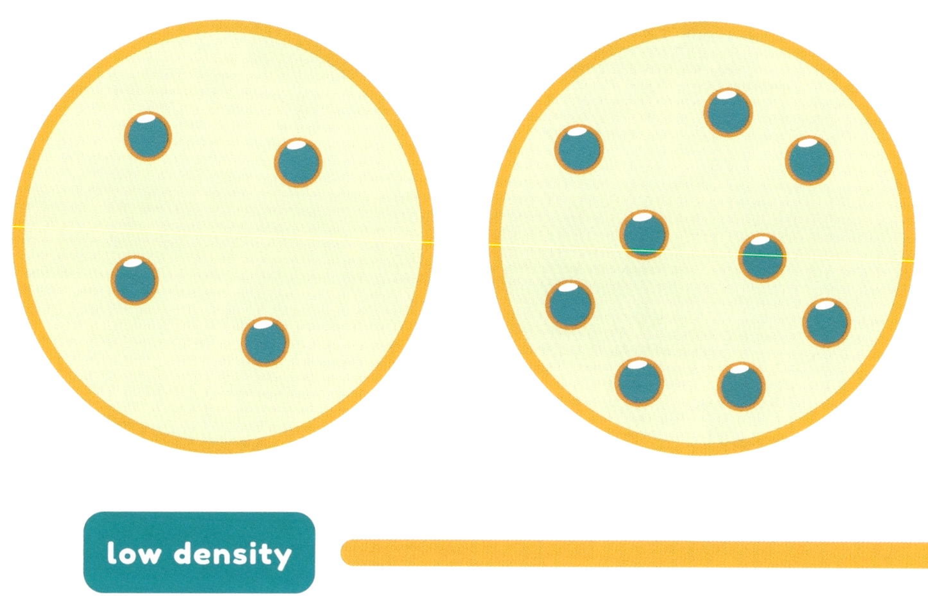

These forces are **determined** by the density of the boat and the density of the water.

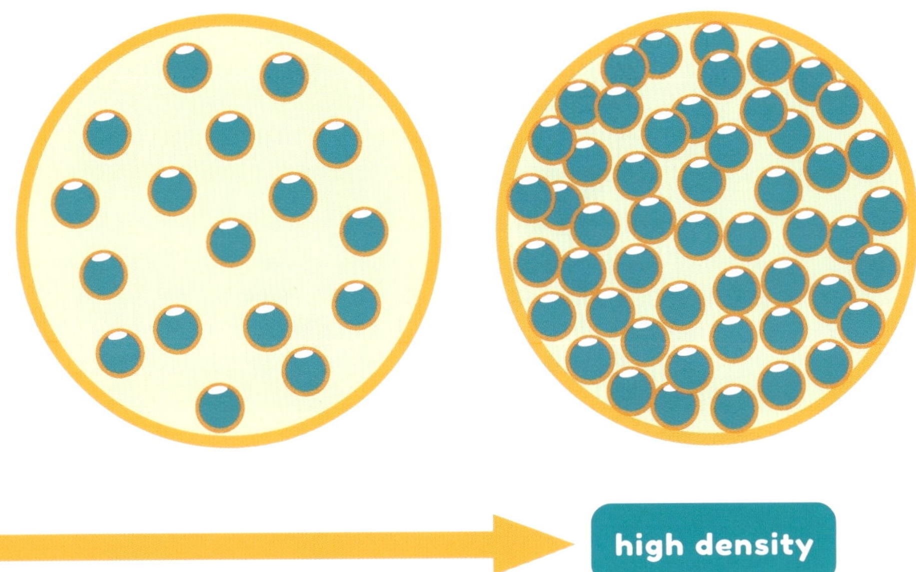

Density is how much space an object takes up compared to how much matter is in the object.

Chapter 3

What Matters?

Matter is anything that takes up space. Everything is made up of matter. The air you breathe, the apple you eat, and even *you* have matter. Mass measures the amount of matter something has.

Learn more here!

A sponge takes up more space than a rock. But it does not have more mass than the rock. This is because the rock is denser. It has more matter inside a smaller space than the sponge.

Even though a boat is big, it doesn't have a lot of matter inside to give it a large mass. There is air in the boat. Air has low density. This keeps the boat's density low.

An inner tube keeps you afloat because it is filled with air. When you're in the tube, your mass spreads across the tube. This lowers your density.

Chapter 4

Above the Surface

When a boat enters water, it's density **displaces** some of the water. The amount of water that moved is equal to the amount of space the boat takes up.

Complete an activity here!

The shape of the boat is also important. A boat has a wide **hull** so its mass can be spread out. The larger the hull is, the more water will be displaced.

This means there will be more force pushing up on the boat.

The density of the boat pushes down against the water. But the water also pushes up on the boat. As long as the boat's mass is less than the mass of the water it displaces, it will stay afloat!

You should always wear a life jacket when you're aboard a boat.

Making Connections

Text-to-Self

Have you ever been on a boat? If so, how big was it? If not, what kind of boat would you first want to go on?

Text-to-Text

Have you read any other books that discuss density and mass? How were they similar to and different from this book?

Text-to-World

How do you think boats changed the world? Write a few sentences explaining your answer.

Glossary

buoyant – having the ability to float.

determine – to decide.

displace – to physically move out of position.

hull – the main part of a boat or ship including the deck, sides, and bottom.

Index

Online Resources

popbooksonline.com

Thanks for reading this Cody Koala book!

Scan this code* and others like it in this book, or visit the website below to make this book pop!

popbooksonline.com/boat-float

*Scanning QR codes requires a web-enabled smart device with a QR code reader app and a camera.